The publishing house tredition has created the series **TREDITION CLASSICS**. It contains classical literature works from over two thousand years. Most of these titles have been out of print and off the bookstore shelves for decades.

The book series is intended to preserve the cultural legacy and to promote the timeless works of classical literature. As a reader of a **TREDITION CLASSICS** book, the reader supports the mission to save many of the amazing works of world literature from oblivion.

The symbol of **TREDITION CLASSICS** is Johannes Gutenberg (1400 – 1468), the inventor of movable type printing.

With the series, tredition intends to make thousands of international literature classics available in printed format again – worldwide.

All books are available at book retailers worldwide in paperback and in hardcover. For more information please visit: www.tredition.com

tredition was established in 2006 by Sandra Latusseck and Soenke Schulz. Based in Hamburg, Germany, tredition offers publishing solutions to authors and publishing houses, combined with worldwide distribution of printed and digital book content. tredition is uniquely positioned to enable authors and publishing houses to create books on their own terms and without conventional manufacturing risks.

For more information please visit: www.tredition.com

The Beginnings of Cheap Steel

Philip W. Bishop

Imprint

This book is part of the TREDITION CLASSICS series.

Author: Philip W. Bishop
Cover design: toepferschumann, Berlin (Germany)

Publisher: tredition GmbH, Hamburg (Germany)
ISBN: 978-3-8491-4735-8

www.tredition.com
www.tredition.de

Copyright:
The content of this book is sourced from the public domain.

The intention of the TREDITION CLASSICS series is to make world literature in the public domain available in printed format. Literary enthusiasts and organizations worldwide have scanned and digitally edited the original texts. tredition has subsequently formatted and redesigned the content into a modern reading layout. Therefore, we cannot guarantee the exact reproduction of the original format of a particular historic edition. Please also note that no modifications have been made to the spelling, therefore it may differ from the orthography used today.

STEEL BEFORE THE 1850's

BESSEMER AND HIS COMPETITORS

ROBERT MUSHET

EBBW VALE AND THE BESSEMER PROCESS

MUSHET AND BESSEMER

WILLIAM KELLY'S AIR-BOILING PROCESS

CONCLUSIONS

THE BEGINNINGS OF CHEAP STEEL

By Philip W. Bishop

Other inventors claimed a part in the invention of the Bessemer process of making steel. Here, the contemporary discussion in the technical press is re-examined to throw light on the relations of these various claimants to the iron and steel industry of their time, as having a possible connection with the antagonism shown by the ironmasters toward Bessemer's ideas.

The Author: *Philip W. Bishop is curator of arts and manufactures, Museum of History and Technology, in the Smithsonian Institution's United States National Museum.*

The development of the world's productive resources during the 19th century, accelerated in general by major innovations in the field of power, transportation, and textiles, was retarded by the occurrence of certain bottlenecks. One of these affected the flow of suitable and economical raw materials to the machine tool and transportation industries: in spite of a rapid growth of iron production, the methods of making steel remained as they were in the previous century; and outputs remained negligible.

In the decade 1855-1865, this situation was completely changed in Great Britain and in Europe generally; and when the United States emerged from the Civil War, that country found itself in a position to take advantage of the European innovations and to start a period of growth which, in the next 50 years, was to establish her as the world's largest producer of steel.

This study reviews the controversy as to the origin of the process which, for more than 35 years[1] provided the greater part of the steel production of the United States. It concerns four men for whom priority of invention in one or more aspects of the process has been claimed.

The process consists in forcing through molten cast iron, held in a vessel called a converter, a stream of cold air under pressure. The combination of the oxygen in the air with the silicon and carbon in the metal raises the temperature of the latter in a spectacular way and after "blowing" for a certain period, eliminates the carbon from the metal. Since steel of various qualities demands the inclusion of from 0.15 to 1.70 percent of carbon, the blow has to be terminated before the elimination of the whole carbon content; or if the carbon content has been eliminated the appropriate percentage of carbon has to be put back. This latter operation is carried out by adding a precise quantity of manganiferous pig-iron (spiegeleisen) or ferro-manganese, the manganese serving to remove the oxygen, which has combined with the iron during the blow.

The controversy which surrounded its development concerned two aspects of the process: The use of the cold air blast to raise the temperature of the molten metal, and the application of manganese to overcome the problem of control of the carbon and oxygen content.

Bessemer, who began his experiments in the making of iron and steel in 1854, secured his first patent in Great Britain in January 1855, and was persuaded to present information about his discovery to a meeting of the British Association for the Advancement of Science held at Cheltenham, Gloucestershire, in August 1856. His title "The Manufacture of Iron without Fuel" was given wide publicity in Great Britain and in the United States. Among those who wrote to the papers to contest Bessemer's theories were several claimants to priority of invention.

Two men claimed that they had anticipated Bessemer in the invention of a method of treating molten metal with air-blasts for the purpose of "purifying" or decarbonizing iron. Both were Americans. Joseph Gilbert Martien, of Newark, New Jersey, who at the time of Bessemer's address was working at the plant of the Ebbw Vale Iron Works, in South Wales, secured a provisional patent a few days before Bessemer obtained one of his series of patents for making cast steel, a circumstance which provided ammunition for those who wished to dispute Bessemer's somewhat spectacular claims. William Kelly, an ironmaster of Eddyville, Kentucky, brought into action by an American report of Bessemer's British Association paper, opposed the granting of a United States patent to Bessemer and substantiated, to the satisfaction of the Commissioner of Patents, his claim to priority in the "air boiling" process.

A third man, this one a Scot resident in England, intervened to claim that he had devised the means whereby Martien's and Bessemer's ideas could be made practical. He was Robert Mushet of Coleford, Gloucestershire, a metallurgist and self-appointed "sage" of the British iron and steel industry who also was associated with the Ebbw Vale Iron Works as a consultant. He, like his American contemporaries, has become established in the public mind as one upon whom Henry Bessemer was dependent for the origin and success of his process. Since Bessemer was the only one of the group to make money from the expansion of the steel industry consequent upon the introduction of the new technique, the suspicion has remained that he exploited the inventions of the others, if indeed he did not steal them.

In this study, based largely upon the contemporary discussion in the technical press, the relation of the four men to each other is reexamined and an attempt is made to place the controversy of 1855-1865 in focus. The necessity for a reappraisal arises from the fact that today's references to the origin of Bessemer steel[2] often contain chronological and other inaccuracies arising in many cases from a dependence on secondary and sometimes unreliable sources. As a result, Kelly's contribution has, perhaps, been overemphasized, with the effect of derogating from the work of another American, Alexander Lyman Holley, who more than any man is entitled to credit for establishing Bessemer steel in America.[3]

Steel Before the 1850's

In spite of a rapid increase in the use of machines and the overwhelming demand for iron products for the expanding railroads, the use of steel had expanded little prior to 1855. The methods of production were still largely those of a century earlier. Slow preparation of the steel by cementation or in crucibles meant a disproportionate consumption of fuel and a resulting high cost. Production in small quantities prevented the adoption of steel in uses which required large initial masses of metal. Steel was, in fact, a luxury product.

The work of Réaumur and, especially, of Huntsman, whose development of cast steel after 1740 secured an international reputation for Sheffield, had established the cementation and crucible processes as the primary source of cast steel, for nearly 100 years. Josiah Marshall Heath's patents of 1839, were the first developments in the direction of cheaper steel, his process leading to a reduction of from 30 to 40 percent in the price of good steel in the Sheffield market.[4] Heath's secret was the addition to the charge of from 1 to 3 percent of carburet of manganese[5] as a deoxidizer. Heath's failure to word his patent so as to cover also his method of producing carburet of manganese led to the effective breakdown of that patent and to the general adoption of his process without payment of license or royalty. In spite of this reduction in the cost of its production, steel remained, until after the midpoint of the century, an insignificant item in the output of the iron and steel industry, being used principally in the manufacture of cutlery and edge tools.

The stimulus towards new methods of making steel and, indeed, of making new steels came curiously enough from outside the established industry, from a man who was not an ironmaster—Henry Bessemer. The way in which Bessemer challenged the trade was itself unusual. There are few cases in which a stranger to an industry has taken the risk of giving a description of a new process in a public forum like a meeting of the British Association for the Advancement of Science. He challenged the trade, not only to attack his theories but to produce evidence from their own plants that they could provide an alternative means of satisfying an emergent demand. Whether or not Bessemer is entitled to claim priority of invention, one can but agree with the ironmaster who said:[6] "Mr. Bessemer has raised such a spirit of enquiry throughout ... the land as must lead to an improved system of manufacture."

Bessemer and his Competitors

Henry Bessemer (1813-1898), an Englishman of French extraction, was the son of a mechanical engineer with a special interest in metallurgy. His environment and his unusual ability to synthesize his observation and experience enabled Bessemer to begin a career of invention by registering his first patent at the age of 25. His active experimenting continued until his death, although the public record of his results ended with a patent issued on the day before his seventieth birthday. A total of 117 British patents[7] bear his name, not all of them, by any means, successful in the sense of producing a substantial income. Curiously, Bessemer's financial stability was assured by the success of an invention he did not patent. This was a process of making bronze powder and gold paint, until the 1830's a secret held in Germany. Bessemer's substitute for an expensive imported product, in the then state of the patent laws, would have failed to give him an adequate reward if he had been unable to keep his process secret. To assure this reward, he had to design, assemble, and organize a plant capable of operation with a minimum of hired labor and with close security control. The fact that he kept the method secret for 40 years, suggests that his machinery[8] (Bessemer describes it as virtually automatic in operation) represented an appreciation of coordinated design greatly in advance of his time. His experience must have directly contributed to his conception of his steel process not as a metallurgical trick but as an industrial process; for when the time came, Bessemer patented his discovery as a process rather than as a formula.

In the light of subsequent developments, it is necessary to consider Bessemer's attitude toward the patent privilege. He describes his secret gold paint as an example of "what the public has had to pay for not being able to give ... security to the inventor" in a situation where the production of the material "could not be identified as having been made by any particular form of mechanism."[9] The inability to obtain a patent over the method of production meant that the disclosure of his formula, necessary for patent specification, would openly invite competitors, including the Germans, to evolve their own techniques. Bessemer concludes:[10]

Had the invention been patented, it would have become public property in fourteen years from the date of the patent, after which period the public would have been able to buy bronze powder at its present [*i.e.*, *ca.* 1890] market price, viz. from two shillings and three pence to two shillings and nine pence per pound. But this important secret was kept for about thirty-five years and the public had to pay excessively high prices for twenty-one years longer than they would have done had the invention become public property in fourteen years, as it would have been if patented. Even this does not represent all the disadvantages resulting from secret manufacture. While every detail of production was a profound secret, there were no improvements made by the outside public in any one of the machines employed during the whole thirty-five years; whereas during the fourteen years, if the invention had been patented, there would, in all probability have been many improved machines invented and many novel features applied to totally different manufactures.

While these words, to some extent, were the rationalizations of an old man, Bessemer's career showed that his philosophy had a practical foundation; and, if this was indeed his belief, the episode explains in large measure Bessemer's later insistence on the legal niceties of the patent procedure. The effect of this will be seen.

Bessemer's intervention in the field of iron and steel was preceded by a period of experiments in the manufacture of glass. Here Bessemer claims to have made glass for the first time in the open hearth of a reverberatory furnace.[11] His work in glass manufacture at least gave him considerable experience in the problems of fusion under high temperatures and provided some support for his later claim that in applying the reverberatory furnace to the manufacture of malleable iron as described in his first patent of January 1855, he had in some manner anticipated the work of C. W. Siemens and Emil Martin.[12]

The general interest in problems of ordnance and armor, stimulated by the Crimean War (1854-1856), was shared by Bessemer, whose ingenuity soon produced a design for a projectile which could provide its own rotation when fired from a smooth-bore gun.[13] Bessemer's failure to interest the British War Office in the idea led him to submit his design to the Emperor Napoleon III. Trials made with the encouragement of the Emperor showed the inadequacy of the cast-iron guns of the period to deal with the heavier shot; and Bessemer was presented with a new problem which, with "the open mind which derived from a limited knowledge of the metallurgy of war," he attacked with impetuosity. Within three weeks of his experiments in France, he had applied for a patent for

"Improvements in the Manufacture of Iron and Steel."[14] This covered the fusion of steel with pig or cast iron and, though this must be regarded as only the first practical step toward the Bessemer process,[15] it was his experiments with the furnace which provided Bessemer with the idea for his later developments.

These were described in his patent dated October 17, 1855 (British patent 2321). This patent is significant to the present study because his application for an American patent, based on similar specifications, led to the interference of William Kelly and to the subsequent denial of the American patent.[16] In British patent 2321 Bessemer proposed to convert his steel in crucibles, arranged in a suitable furnace and each having a vertical tuyère, through which air under pressure was forced through the molten metal. As Dredge[17] points out, Bessemer's association of the air blast with the increase in the temperature of the metal "showed his appreciation of the end in view, and the general way of attaining it, though his mechanical details were still crude and imperfect."

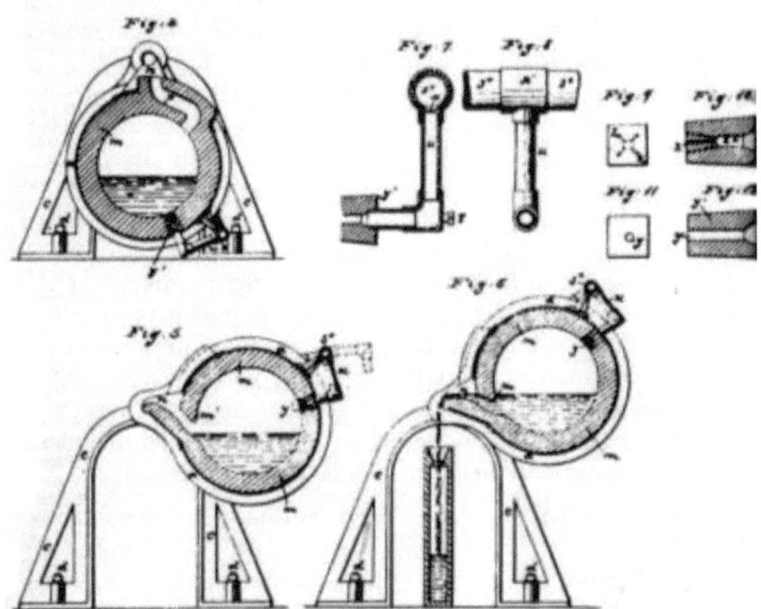

Figure 1.—Bessemer's Design for a Converter, as Shown in U.S. Patent 16082. This patent, dated November 11, 1856, corresponds with British patent 356, dated February 12, 1856. The more familiar design of converter appeared first in British patent 578, March 1, 1860. The contrast with Kelly's schematic drawing in Fig. 2 (p. 42) is noticeable.

Experiments were continued and several more British patents were applied for before Bessemer made his appearance before the British Association on August 13, 1856.[18] Bessemer described his first converter and its operation in some detail. Although he was soon to realize that he "too readily allowed myself to bring my inventions under public notice,"[19] Bessemer had now thrown out a challenge which eventually had to be taken up, regardless of the strength of the vested interests involved. The provocation came from his claims that the product of the first stage of the conversion was the equivalent of charcoal iron, the processes following the smelting being conducted without contact with, or the use of, any mineral fuel; and that further blowing could be used to produce any quality of metal, that is, a steel with any desired percentage of carbon. Yet, the principal irritant to the complacency of the ironmaster must have been Bessemer's attack on an industry which had gone on increasing the size of its smelting furnaces, thus improving the uniformity of its pig-iron, without modifying the puddling process, which at best could handle no more than 400 to 500 pounds of iron at a time, divided into the "homeopathic doses" of 70 or 80 pounds capable of being handled by human labor.[20] Bessemer's claim to "do" 800 pounds of metal in 30 minutes against the puddling furnace's output of 500 pounds in two hours was calculated to arouse the opposition of those who feared the loss of capital invested in puddling furnaces and of those who suspected that their jobs might be in jeopardy. The ensuing criticism of Bessemer has to be interpreted, therefore, with this in mind; not by any means was it entirely based on objective consideration of the method or the product.[21]

Within a month of his address, Bessemer had sold licenses to several ironmasters (outside Sheffield) and so provided himself with capital with which to continue his development work; but he refused to sell his patents outright to the Ebbw Vale Iron Works and by this action, as will be seen, he created an enemy for himself.

The three years between 1856 and 1859, when Bessemer opened his own steel works in Sheffield, were occupied in tracing the causes of his initial difficulties. There was continued controversy in the technical press. Bessemer (unless he used a *nom-de-plume*) took no part in it and remained silent until he made another public appearance before the Institution of Civil Engineers in London (May 1859). By this time Bessemer's process was accepted as a practical one, and the claims of Robert Mushet to share in his achievement was becoming clamorous.

Robert Mushet

Robert (Forester) Mushet (1811-1891), born in the Forest of Dean, Gloucestershire, of a Scots father (David, 1772-1847) himself a noted contributor to the metallurgy of iron and steel, is, like the American William Kelly, considered by many to have been a victim of Bessemer's astuteness—or villainy. Because of Robert Mushet's preference for the quiet of Coleford, many important facts about his career are lacking; but even if his physical life was that of a recluse, his frequent and verbose contributions to the correspondence columns of the technical press made him well-known to the iron trade. It is from these letters that he must be judged.

In view of his propensity to intervene pontifically in every discussion concerning the manufacture of iron and steel, it is somewhat surprising that he refrained from comment on Bessemer's British Association address of August 1856 for more than fourteen months. The debate was opened over the signature of his brother David who shared the family facility with the pen.[22] Recognizing Bessemer's invention as a "congruous appendage to [the] now highly developed powers of the blast furnace" which he describes as "too convenient, too powerful and too capable of further development to be superseded by any retrograde process," David Mushet greeted Bessemer's discovery as "one of the greatest operations ever devised in metallurgy."[23] A month later, however, David Mushet had so modified his opinion of Bessemer as to come to the conclusion that the latter "must indeed be classed with the most unfortunate inventors." He gave as his reason for this turnabout his discovery that Joseph Martien had demonstrated his process of "purifying" metal

15

successfully and had indeed been granted a provisional patent a month before Bessemer. The sharp practice of Martien's patent lawyer, Mushet claimed, had deprived him of an opportunity of proving priority of invention against Bessemer. Mushet was convinced that Martien's was the first in the field.[24]

Robert Mushet's campaign on behalf of his own claims to have made the Bessemer process effective was introduced in October 1857, two years after the beginning of Bessemer's experiment and after one year of silence on Bessemer's part. Writing as "Sideros"[25] he gave credit to Martien for "the great discovery that pig-iron can, whilst in the fluid state, be purified ... by forcing currents of air under it ...," though Martien had failed to observe the use of temperature by the "deflation of the iron itself"; and for discovering that—

when the carbon has been all, or nearly all, dissipated, the temperature increases to an almost inconceivable extent, so that the mass, when containing only as much carbon as is requisite to constitute with it cast steel ... still retains a perfect degree of fluidity.

This, says "Sideros," was no new observation; "it had been before the metallurgical world, both practical and scientific, for centuries," but Bessemer was the first to show that this generation of heat could be attained by blowing cold air through the melted iron. Mushet goes on to show, however, that the steel thus produced by Bessemer was not commercially valuable because the sulphur and phosphorous remained, and the dispersion of oxide of iron through the mass "imported to it the inveterate hot-short quality which no subsequent operation could expel." "Sideros" concludes that Bessemer's discovery was "at least for a time" now shelved and arrested in its progress; and it had been left "to an individual of the name of Mushet" to show that if "fluid metallic manganese" were combined with the fluid Bessemer iron, the portion of manganese thus alloyed would unite with the oxygen of the oxide and pass off as slag, removing the hot-short quality of the iron. Robert Mushet had demonstrated his product to "Sideros" and had patented his discovery, though "not one print, literary or scientific, had condescended to notice it."

"Sideros" viewed Mushet's discovery as a "spark amongst dry faggots that will one day light up a blaze which will astonish the world when the unfortunate inventor can no longer reap the fruits of his life-long toil and unflinching perseverance." In an ensuing letter he[26] summed up the situation as he saw it:

Nothing that Mr. Mushet can hereafter invent can entitle him to the merit of Mr. Bessemer's great discovery ... and ... nothing that Mr. Bessemer may hereafter patent can deprive Mr. Robert Mushet of having been the first to remove the obstacles to the success of Mr. Bessemer's process.

Bessemer still did not intervene in the newspaper discussion; nor had he had any serious supporters, at least in the early stage.[27]

Publication in the *Mining Journal* of a list of Mushet's patents,[28] evidently in response to Sideros' complaint, now presented Bessemer with notice of Robert Mushet's activity, even if he had not already observed his claims as they were presented to the Patent Office. Mushet, said the *Mining Journal* —

appears to intend to carry on his researches from the point where Mr. J. G. Martien left off and is proceeding on the Bessemer plan of patenting each idea as it occurs to his imaginative brain. He proposes to make both iron and steel but does not appear to have quite decided as to the course of action ... to accomplish his object, and therefore claims various processes, some of which are never likely to realize the inventor's expectations, although decidedly novel, whilst others are but slight modification of inventions which have already been tried and failed.

The contemporary attitude is reflected in another comment by the *Mining Journal*:[29]

Although the application of chemical knowledge to the manufacture of malleable iron cannot fail to produce beneficial results, the quality of the metal depends more upon the mechanical than the chemical processes.... Without wishing in any way to discourage the iron chemists, we have no hesitation in giving this as our opinion which we shall maintain until the contrary be actually proved. With regard to steel, there may be a large field for chemical research ... however, we believe that unless the iron be of a nature adapted for the manufacture of steel by ordinary processes, the purely chemical inventions will only give a metal of a very uniform quality.

Another correspondent, William Green, was of the opinion that Mushet's "new compounds and alloys," promised well as an auxiliary to the Bessemer process but that "the evil which it was intended to remove was more visionary than real." Bessemer's chief difficulty was the phosphorus, not the oxide of iron "as Mr. Mushet assumes." This, Bessemer no doubt would deal with in due course, but meanwhile he did well "to concentrate his energies upon the steel operations," after which he would have time to tackle "the difficulties which have so far retarded the iron operations."[30]

Mushet[31] claims to have taken out his patent of September 22, 1856, covering the famous "triple compound," after he —

had fully ascertained, upon the ordinary scale of manufacture that air-purified cast-iron, when treated as set forth in my specifications, would afford tough malleable iron ... I found, however, that the remelting of the coke pig-iron, in contact with coke fuel, hardened the iron too much, and it became evident that an air-furnace was more proper for my purpose ... [the difficulties] arose, not from any defect in my process, but were owing to the small quantity of the metal operated upon and the imperfect arrangement of the purifying vessel, which ought to be so constituted that it may be turned upon an axis, the blast taken off, the alloy added and the steel poured out through a spout ... *Such a purifying vessel Mr. Bessemer has delineated in one of his patents.*

Mushet also claimed to have designed his own "purifying and mixing" furnace, of 20-ton capacity, which he had submitted to the Ebbw Vale Iron Works "many months ago," without comment from them. There is an intriguing reference to the painful subject of two patents not proceeded with, and not discussed "in the avaricious hope that the parties connected with the patents will make me honorable amends ... these patents were suppressed without my knowledge or consent." Lest his qualifications should be questioned, Mushet concludes:

I do not profess to be an iron chemist, but I have undoubtedly made more experiments upon the subject of iron and steel than any man now living and I am thereby enabled to say that all I know is but little in comparison with what has yet to be discovered.

So began Mushet's claim to have solved Bessemer's problem, a claim which was to fill the correspondence columns of the engineering journals for the next ten years. Interpretation of this correspondence is made difficult by our ignorance of the facts concerning the control of Mushet's patents. These have to be pieced together from his scattered references to the subject.

His experiments were conducted, at least nearly up to the close of the year 1856, with the cooperation of Thomas Brown of the Ebbw Vale Iron Works.[32] The price of this assistance was apparently half interest in Mushet's patents, though for reasons which Mushet does not explain the deed prepared to effect the transfer was never executed.[33] Mushet continued, however, to regard the patents as "wholly my own, though at the same time, I am bound in honor to take no unfair advantage of the non-execution of that deed." A possible explanation of this situation may be found in Ebbw Vale's activities in connection with Martien and Bessemer, as well as with an Austrian inventor, Uchatius.

Ebbw Vale and the Bessemer Process

After his British Association address in August 1856, Bessemer had received applications from several ironmasters for licenses, which were issued in return for a down payment and a nominal royalty of 25 pence per ton. Among those who started negotiations was Mr. Thomas Brown of Ebbw Vale Iron Works, one of the largest of the South Wales plants. He proposed, however, instead of a license, an outright purchase of Bessemer's patents for £50,000. Bessemer refused to sell, and according to his[34] account—

intense disappointment and anger quite got the better of [Brown] and for the moment he could not realize the fact of my refusal.... [He then] left me very abruptly, saying in an irritated tone ... "I'll make you see the matter differently yet" and slammed the door after him.

David Mushet's advocacy of Martien's claim to priority over Bessemer has already been noticed (p. 33). From him we learn[35] that Martien's experiments leading to his patent of September 15, 1855, had been carried out at the Ebbw Vale Works in South Wales, where he engaged in "perfecting the Renton process."[36] Martien's own process consisted in passing air through metal as it was run in a trough from the furnace and before it passed into the puddling furnace.

It is known that Martien's patent was in the hands of the Ebbw Vale Iron Works by March 1857.[37] This fact must be added to our knowledge that Mushet's patent of September 22, 1856 was drawn up with a specific reference to the application of his "triple compound" to "iron ... purified by the action of air, in the manner invented by Joseph Gilbert Martien,"[38] and that this and his other manganese patents were under the effective control of Ebbw Vale. It seems a reasonable deduction from these circumstances that Brown's offer to buy out Bessemer and his subsequent threat were the consequences of a determination by Ebbw Vale to attack Bessemer by means of patent infringement suits.

Some aspects of the Ebbw Vale situation are not yet explained. Martien came to South Wales from Newark, New Jersey, where he had been manager of Renton's Patent Semi-Bituminous Coal Furnace, owned by James Quimby, and where he had something to do with the installation of Renton's first furnace in 1854. The first furnace was unsuccessful.[39] Martien next appears in Britain, at the Ebbw Vale Iron Works. No information is available as to whether Martien's own furnace was actually installed at Ebbw Vale, although as noted above, David Mushet claims to have been invited to see it there.

Martien secured an American patent for his process in 1857 and to file his application appears to have gone to the United States, where he remained at least until October 1858.[40] He seems to have taken the opportunity to apply for another patent for a furnace similar to that of James Renton. This led to interferences proceedings in which Martien showed that he had worked on this furnace at Bridgend, Glamorganshire (one of the Ebbw Vale plants), improving Renton's design by increasing the number of "deoxydizing tubes." This variation in Renton's design was held not patentable, and in any case Renton's firm was able to show that they had successfully installed the furnace at Newark in 1852-1853, while Martien could not satisfy the Commissioner that his installation had been made before September 1854. Priority was therefore awarded to Quimby, Brown, Renton, and Creswell.[41]

Since Renton had not patented his furnace in Great Britain, Martien's use of his earlier knowledge of Renton's work and of his experience at Bridgend in an attempt to upset Renton's priority is a curious and at present unexplainable episode. Perhaps the early records of the Ebbw Vale Iron Works, if they exist, will show whether this episode was in some way linked to the firm's optimistic combination of the British patents of Martien and Mushet.

That Ebbw Vale exerted every effort to find an alternative to Bessemer's process is suggested, also, by their purchase in 1856 of the British rights to the Uchatius process, invented by an Austrian Army officer. The provisional patent specifications, dated October 1, 1855, showed that Uchatius proposed to make cast steel directly from pig-iron by melting granulated pig-iron in a crucible with pulverized "sparry iron" (siderite) and fine clay or with gray oxide of manganese, which would determine the amount of carbon combining with the iron. This process, which was to prove commercially successful in Great Britain and in Sweden but was not used in America,[42] appeared to Ebbw Vale to be something from which, "we can have steel produced at the price proposed by Mr. Bessemer, notwithstanding the failure of his process to fulfil the promise."[43]

So far as is known only one direct attempt was made, presumably instigated by Ebbw Vale, to enforce their patents against Bessemer, who records[44] a visit by Mushet's agent some two or three months before a renewal fee on Mushet's basic manganese patents became payable in 1859. Bessemer "entirely repudiated" Mushet's patents and offered to perform his operations in the presence of Mushet's lawyers and witnesses at the Sheffield Works so that a prosecution for infringement "would be a very simple matter." That, he says, was the last heard from the agent or from Mushet on the subject.[45] The renewal fee was not paid and the patents were therefore abandoned by Ebbw Vale and their associates, a fact which did not come to Mushet's knowledge until 1861, when he himself declared that the patent "was never in my hands at all [so] that I could not enforce it."[46]

Further support for the thesis that Ebbw Vale's policy was in part dictated by a desire to make Bessemer "see the matter differently" is to be found in the climatic episode. Work on Martien's patents had not been abandoned and in 1861 certain patents were taken out by George Parry, Ebbw Vale's furnace manager. These, represented as improvements of Martien's designs, were regarded by Bessemer as clear infringements of his own patents.[47] When it came to Bessemer's knowledge that Ebbw Vale was proposing to "go to the public" for additional capital with which to finance, in part, a large scale working of Parry's process, he threatened the financial promoter with injunctions and succeeded in opening negotiations for a settlement. All the patents "which had been for years suspended" over Bessemer were turned over to him for £30,000. Ebbw Vale, thereupon, issued their prospectus[48] with the significant statement that the directors "have agreed for a license for the manufacture of steel by the Bessemer process which, from the peculiar resources they possess, they will be enabled to produce in very large quantities...." So Bessemer became the owner of the Martien and Parry patents. Mushet's basic patents no longer existed.

Mushet and Bessemer

That Mushet was "used" by Ebbw Vale against Bessemer is, perhaps, only an assumption; but that he was badly treated by Ebbw Vale is subject to no doubt. Mushet's business capacity was small but it is difficult to believe that he could have been so foolish as to assign an interest in his patents to Ebbw Vale without in some way insuring his right of consultation about their disposition. He claims that even in the drafting of his specifications he was obliged to follow die demands of Ebbw Vale, which firm, believing, "on the advice of Mr. Hindmarsh, the most eminent patent counsel of the day,"[49] that Martien's patent outranked Bessemer's, insisted that Mushet link his process to Martien's. This, as late as 1861, Mushet believed to be in effective operation.[50] His later repudiation of the process as an absurd and impracticable patent process "possessing neither value nor utility"[51] may more truly represent his opinion, especially as, when he wrote his 1861 comment, he still did not know of the disappearance of his patents.

Mushet's boast[52] that he had never been into an ironworks other than his own in Coleford is a clue to the interpretation of his behavior in general and also of his frequent presumptuous claims. When, for instance, the development of the Uchatius process was publicized, he gave his opinion[53] that the process was a useless one and had been patented before Uchatius "understood its nature"; yet later[54] he could claim that the process was "in fact, my own invention and I had made and sold the steel thus produced for some years previously to the date of Captain Uchatius' patent". Moreover, he claims to have instructed Uchatius' agents in its operation! He may, at this later date, have recalled his challenge (the first of many such) in which he offered Uchatius' agent in England to pay a monetary penalty if he could not show a superior method of producing "sound serviceable cast steel from British coke pig-iron, *on the stomic plan* and without any mixture of clay, oxide of manganese or any of these pot destroying ingredients."[55]

It was David Mushet (or Robert, using his brother's name)[56] who accused Bessemer, or rather his patent agent, Carpmael, of sharp practice in connection with Martien's specification, an allegation later supported by Martien's first patent agent, Avery.[57] The story was that for the drafting of his final specification, Martien, presumably with the advice of the Ebbw Vale Iron Works, consulted the same Carpmael, as "the leading man" in the field. The latter advised that the provisional specification restricted Martien to the application of his method to iron flowing in a channel or gutter from the blast furnace, and so prevented him from applying his aeration principle in any kind of receptacle. In effect, Carpmael was acting unprofessionally by giving Bessemer the prior claim to the use of a receptacle. According to Mushet, Martien had in fact "actually and publicly proved" his process in a receptacle and not in a gutter, so that his claim to priority could be maintained on the basis of the provisional specification.

This, like other Mushet allegations, was ignored by Bessemer, and probably with good reason. At any rate, Martien's American patent is in terms similar to those of the British specification; he or his advisers seem to have attached no significance to the distinction between a gutter and a receptacle.

Mushet's claim to have afforded Bessemer the means of making his own process useful is still subject to debate. Unfortunately, documentation of the case is almost wholly one sided, since his biggest publicizer was Mushet himself. An occasional editorial in the technical press and a few replies to Mushet's "lucubrations" are all the material which exists, apart from Bessemer's own story.

Mushet and at least five other men patented the use of manganese in steel making in 1856; his own provisional specification was filed within a month of the publication of Bessemer's British Association address in August 1856. So it is strange that Robert Mushet did not until more than a year later join in the controversy which followed that address.[58] In one of his early letters he claims to have made of "his" steel a bridge rail of 750 pounds weight; although his brother insists that he saw the same rail in the Ebbw Vale offices in London in the spring of 1857, when it was presented as a specimen of Uchatius steel![59] Robert Mushet's indignant "advertisement" of January 5, 1858,[60] reiterating his parentage of this sample, also claimed a double-headed steel rail "made by me under another of my patent processes," and sent to Derby to be laid down there to be "subjected to intense vertricular triturations." Mushet's description of the preparation of this ingot[61] shows that it was derived from "Bessemer scrap" made by Ebbw Vale in the first unsuccessful attempts of that firm to simulate the Bessemer process. This scrap Mushet had remelted in pots with spiegel in the proportions of 44 pounds of scrap to 3 of melted spiegel. It was his claim that the rail was rolled direct from the ingot, something Bessemer himself could not do at that time.

This was the beginning of a series of claims by Mushet as to his essential contributions to Bessemer's invention. The silence of the latter during this period is impressive, for according to Bessemer's own account[62] his British Association address was premature, and although the sale of licenses actually provided him with working funds, the impatience of those experimenting with the process and the flood of competing "inventions" all embarrassed him at the most critical stage of this development of the process: "It was, however, no use for me to argue the matter in the press. All that I could say would be mere talk and I felt that action was necessary, and not words."[63]

Action took the form of continued experiments and, by the end of 1857, a decision to build his own plant at Sheffield.[64] An important collateral development resulted from the visit to London in May 1857 of G. F. Goransson of Gefle, Sweden. Using Bessemer equipment, Goransson began trials of the process in November 1857 and by October 1858 was able to report: "Our firm has now entirely given up the manufacture of bar iron, and our blast furnaces and tilt mills are now wholly employed in making steel by the Bessemer process, which may, therefore, be now considered an accomplished commercial fact."[65]

Goransson was later to claim considerable improvements on the method of introducing the blast, and, in consequence, the first effective demonstration of the Bessemer method[66] — this at a time when Bessemer was still remelting the product of his converter in crucibles, after granulating the steel in water. If Mushet is to be believed, this success of Goransson's was wholly due to his ore being "totally free from phosphorous and sulphur."[67] However, Bessemer's own progress was substantial, for his Sheffield works were reported as being in active operation in April 1859, and a price for his engineers' tool and spindle steel was included in the *Mining Journal* "Mining Market" weekly quotations for the first time[68] on June 4, 1859.

In May 1859 Bessemer gave a paper, his first public pronouncement since August 1856, before the Institution of Civil Engineers.[69] The early process, he admitted, had led to failure because the process had not reduced the quantity of sulphur and phosphorous, but his account is vague as to the manner in which he dealt with this problem:

Steam and pure hydrogen gas were tried, with more or less success in the removal of sulphur, and various flues, composed chiefly of silicates of the oxide of iron and manganese were brought in contact with the fluid metal, during the process and the quantity of phosphorous was thereby reduced.

But the clear implication is that the commercial operation at Sheffield was based on the use of the best Swedish pig iron and the hematite pig from Workington. The use of manganese as standard practice at this time is not referred to,[70] but the rotary converter and the use of ganister linings are mentioned for the first time.

Mushet had, with some intuition, found opportunity to reassert his contributions to Bessemer a few days before this address, describing his process as perhaps lacking "the extraordinary merit of Mr. Bessemer," being "merely a vigorous offshoot proceeding from that great discovery; but, combined with Mr. Bessemer's process, it places within the reach of every iron manufacturer to produce cast steel at the same cost for which he can now make his best iron."[71]

One of Mushet's replies to the paper itself took the form of the announcement of his provisional patent for the use of his triple compound which, in the opinion of *The Mining Journal* appeared to be "but a very slight modification of several of Mr. Bessemer's inventions." Another half dozen patents appeared within two months, "so that it is apparent that Mr. Mushet's failure to make the public appreciate his theories has not injured his inventive faculties."[72] These patents include, besides variations on his "triple compound" theme, his important patent on the use of tungsten for cutting tools, later to be known as Mushet steel.[73]

Mushet's formal pronouncement on Bessemer's paper, dated June 28, 1859, is perhaps his most intelligible communication on the subject. He alone "from the first consistently advocated the merits and pointed out the defects of the Bessemer process," and within a few days of the British Association address he had shown Ebbw Vale "where the defect would be found and what would remedy" it. It was not, in fact, the presence of one-tenth of a percent of sulphur or phosphorous which affected the result if the Bessemer process were combined with his process by adding a triple compound of iron, carbon, and manganese to the pig. "There never was a bar of first-rate cast steel made by the Bessemer process alone"; (and that included Goransson's product) "and there never can be, but a cheap kind of steel applicable to several purposes may be thus produced." After emphasizing the uniqueness of his attempt to make Bessemer's process successful, he asserts:[74]

In short, I merely availed myself of a great metallurgical fact, *which has been for years* before the eyes of the metallurgical world, namely that the presence of metallic manganese in iron and steel conferred upon both an amount of toughness either when cold or when heated, which the presence at the same time of a notable amount of sulphur and phosphorous could not overcome.

The succeeding years were enlivened, one by one, by some controversy in which Mushet invoked the shadow of his late father as support for some pronouncement, or "edict," as some said, on the subject of making iron and steel. In 1860, on the question of suitable metal for artillery, later to be the subject of high controversy among the leading experts of the day, Mushet found a ready solution in his own gun metal. This he had developed fifteen years before. It was of a tensile strength better even than that of Krupp of Essen who was then specializing in the making of large blocks of cast steel for heavy forgings, and particularly for guns. Indeed, he was able publicly to challenge Krupp to produce a cast gun metal or cast steel to stand test against his.[75] A year later his attack on the distinguished French metallurgist Fremy, whom he describes as an "ass" for his interest in the so-called cyanogen process of steel making, did little to enhance his reputation, whatever the scientific justification for his attack. His attitude toward the use of New Zealand (Taranaki) metalliferous sand, which he had previously favored and then condemned in such a way as to "injure a project he can no longer control,"[76] was another example of a public behavior evidently resented.

By mid-1861, on the other hand, Bessemer was beginning to meet with increasing respect from the trade. The Society of Engineers received a dispassionate account of the achievement at the Sheffield Works from E. Riley, whose firm (Dowlais) was among the earlier and disappointed licensees of the process.[77] In August 1861, five years after the ill-fated address before the British Association, the Institution of Mechanical Engineers, meeting in Sheffield, the center of the British steel trade, heard papers from Bessemer and from John Brown, a famous ironmaster. The latter described the making of Bessemer rails, the product which above all was to absorb the Bessemer plants in America after 1865. After the meeting, the engineers visited Bessemer's works; and later it was reported,[78] "at Messrs. John Brown and Company's works, the Bessemer process was repeated on a still larger scale and a heavy armor plate rolled in the presence of some 250 visitors...."

These proceedings invited Robert Mushet's intervention. Still under the impression that his patent was still alive and, with Martien's, in the "able hands" of the Ebbw Vale Iron Company, he condemned Bessemer for his "lack of grace" to do him justice, and

27

took the occasion to indict the patent system which denied him and Martien the fruits of their labors.[79]

The Engineer found Mushet's position untenable on the very grounds he was pleading—that patents should not be issued to different men at different times for the same thing; and showed that Bessemer in his patents of January 4, 1856, and later, had clearly anticipated Mushet. In a subsequent article, *The Engineer* disposed of Martien's and Mushet's claims with a certain finality. The Ebbw Vale Iron Works had spent £7,000 trying to carry out the Martien process and it was unlikely that they would have allowed Bessemer to infringe upon that patent if they had any grounds for a case. Bessemer was not imitating Mushet. The latter's "triple compound" required manganese pig-iron (with a content of 2 to 5 percent of manganese) at £13 per ton while Bessemer used an oxide of manganese (at a 50 percent concentration): at £7 per ton.

The alloy of manganese and other materials now used in the atmospheric process contains 50 percent of manganese a proportion which could never be obtained from the blast furnace, owing to the highly oxidisable nature of that metal. And it is absolutely necessary, in order to apply any useful alloy of iron, carbon and manganese, in the manufacture of malleable iron and very soft steel that the manganese should be largely in excess of the carbon present.[80]

Sufficient answer to Mushet was at any rate available in the fact that many hundreds of tons of excellent "Bessemer metal" made without any mixture of manganese or spiegeleisen in any form were in successful use. And, moreover, spiegeleisen was not a discovery of Robert Mushet or an exclusive product of Germany since it had been made for twenty years at least from Tow Law (Durham) ores. If Bessemer had refused Mushet a license (and this was an admitted fact), Bessemer's refusal must have been made in self-defense:

Mr. Mushet having set up a number of claims for "improvements" upon which claims, we have a right to suppose, he was preparing to take toll from Mr. Bessemer, but which claims, the latter gentleman discovered, in time, were worthless and accordingly declined any negotiations with the individual making them.[81]

Mushet's claims were by this time rarely supported in the periodicals. One interesting article in his favor came in 1864 from a source of special interest to the American situation. Mushet's American patent[82] had been bought by an American group interested in the Kelly process at about this time,[83] and Bessemer's American rights had also been sold to an American group that included Alexander Lyman Holley,[84] who had long been associated with Zerah Colburn, another American engineer. Colburn, who subsequently (1866) established the London periodical *Engineering* and is regarded as one of the founders of engineering journalism, was from 1862 onward a frequent contributor to other trade papers in London. Colburn's article of 1864[85] seems to have been of some importance to Mushet, who, in the prospectus of the Titanic Steel and Iron Company, Ltd., issued soon after, brazenly asserted[86] that, "by the process of Mr. Mushet *especially when in combination with the Bessemer process*, steel as good as Swedish steel" would be produced at £6 per ton. Mushet may have intended to invite a patent action, but evidently Bessemer could now more than ever afford to ignore the "sage of Coleford."

The year 1865 saw Mushet less provocative and more appealing; as for instance: "It was no fault of Mr. Bessemer's that my patent was lost, but he ought to acknowledge his obligations to me in a manly, straightforward manner and this would stamp him as a great man as well as a great inventor."[87]

But Bessemer evidently remained convinced of the security of his own patent position. In an address before the British Association at Birmingham in September 1865 he made his first public reply to Mushet.[88] In his long series of patents Mushet had attempted to secure—

almost every conceivable mode of introducing manganese into the metal.... Manganese and its compounds were so claimed under all imaginable conditions that if this series of patents could have been sustained in law, it would have been utterly impossible for [me] to have employed manganese with steel made by his process, although it was considered by the trade to be impossible to make steel from coke-made iron without it.

The failure of those who controlled Mushet's batch of patents to renew them at the end of three years, Bessemer ascribed to the low public estimation to which Mushet's process had sunk in 1859, and he had therefore, "used without scruple any of these numerous patents for manganese without feeling an overwhelming sense of obligation to the patentee." He was now using ferromanganese made in Glasgow. Another alloy, consisting of 60 to 80 percent of metallic manganese was also available to him from Germany.

This renewed publicity brought forth no immediate reply from Mushet, but a year later he was invited to read a paper before the British Association. A report on the meeting stated that in his paper he repeated his oft-told story, and that "he still thought that the accident (of the non-payment of the patent stamp duties) ought not to debar him from receiving the reward to which he was justly entitled." Bessemer, who was present, reiterated his constant willingness to submit the matter to the courts of law, but pointed out that Mushet had not accepted the challenge.[89]

Three months later, in December 1866, Mushet's daughter called on Bessemer and asked his help to prevent the loss of their home: "They tell me you use my father's inventions and are indebted to him for your success." Bessemer replied characteristically:

I use what your father has no right to claim; and if he had the legal position you seem to suppose, he could stop my business by an injunction tomorrow and get many thousands of pounds compensation for my infringement of his rights. The only result which followed from your father taking out his patents was that they pointed out to me some rights which I already possessed, but of which I was not availing myself. Thus he did me some service and even for this unintentional service, I cannot live in a state of indebtedness....

With that he gave Miss Mushet money to cover a debt for which distraint was threatened.[90] Soon after this action, Bessemer made Mushet a "small allowance" of £300 a year. Bessemer's reasons for making this payment, he describes as follows: "There was a strong desire on my part to make him (Mushet) my debtor rather than the reverse, and the payment had other advantages: the press at that time was violently attacking my patent and there was the chance that if any of my licensees were thus induced to resist my claims, all the rest might follow the example."[91]

Mushet's Titanic Steel and Iron Company was liquidated in 1871 and its principal asset, "R. Mushet's special steel," that is, his tungsten alloy tool metal, was taken over by the Sheffield firm of Samuel Osborn and Company. The royalties from this, with Bessemer's pension seem to have left Mushet in a reasonably comfortable condition until his death in 1891;[92] but even the award of the Bessemer medal by the Iron and Steel Institute in 1876 failed to remove the conviction that he had been badly treated. One would like to know more about the politics which preceded the award of the trade's highest honor. Bessemer at any rate was persuaded to approve of the presentation and attended the meeting. Mushet himself did not accept the invitation, "as I may probably not be then alive."[93] The President of the Institute emphasized the present good relations between Mushet and Bessemer and the latter recorded that the hatchet had "long since" been buried. Yet Mushet continued to brood over the injustice done to him and eventually recorded his story of the rise and progress of the "Bessemer-Mushet" process in a pamphlet[94] written apparently without reference to his earlier statements and so committing himself to many inconsistencies.

William Kelly's "Air-boiling" Process

An account of Bessemer's address to the British Association was published in the *Scientific American* on September 13, 1856.[95] On September 16, 1856, Martien filed application for a U.S. patent on his furnace and Mushet for one on the application of his triple compound to cast iron "purified or decarbonized by the action of air blown or forced into ... its particles while it is in a molten ... state."[96] Mushet, by this time, had apparently decided to generalize the application of his compound instead of citing its use in conjunction with Martien's process, or, as he put it, he had been obliged to do for his English specification by the Ebbw Vale Iron Works.

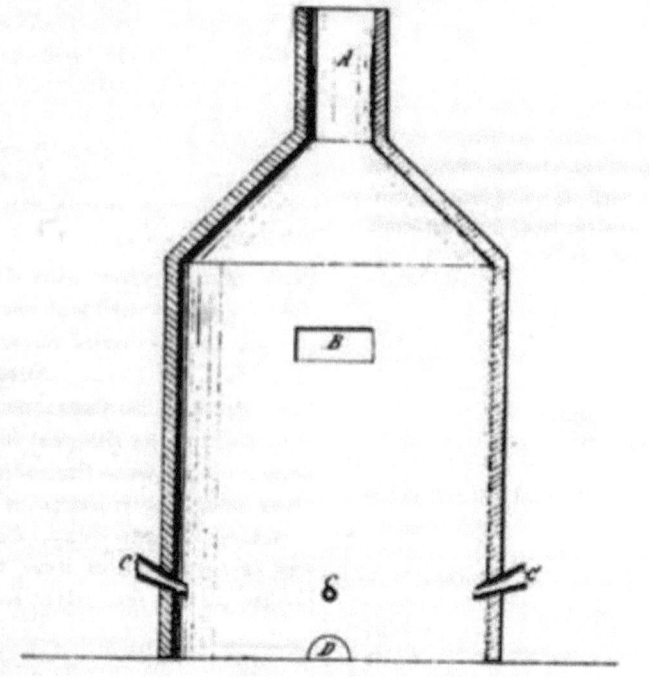

Figure 2.—Only Known Design for Kelly's Air-Boiling Furnace, From U.S. Patent 17628. *A* is "the flue to carry off the carbonic gas formed in decarbonizing the iron," *B* is the port through which the charge of fluid iron is received, *C* and *C'* are the tuyères, and *D* is the tap hole for letting out the refined metal.

The discussion in the *Scientific American*, which was mostly concerned with Martien's claim to priority, soon evoked a letter from William Kelly. Writing under date of September 30, 1856, from the Suwanee Iron Works, Eddyville, Kentucky, he claimed to have started "a series of experiments" in November 1851 which had been witnessed by hundreds of persons and "discussed amongst the ironmasters, etc., of this section, all of whom are perfectly familiar with the whole principle ... as discovered by me nearly five years ago." A number of English puddlers had visited him to see his new process. "Several of them have since returned to England and may have spoken of my invention there." Kelly expected "shortly to have the invention perfected and bring it before the public."[97]

Bessemer's application for an American patent was granted during the week ending November 18, 1856, and Kelly began his interference proceedings sometime before January 1857.[98]

Kelly's witnesses were almost wholly from the ranks of employees or former employees. The only exception was Dr. Alfred H. Champion, a physician of Eddyville. Dr. Champion describes a meeting in the fall of 1851 with "two or three practical Ironmasters and others" at which Kelly described his process and invited all present to see it in operation. He stated:

The company present all differed in opinion from Mr. Kelly and appealed to me as a chemist in confirmation of their doubts. I at once decided that Mr. Kelly was correct in his Theory and then went on to explain the received opinion of chemists a century ago on this subject, and the present received opinion which was in direct confirmation of the novel theory of Mr. Kelly. I also mentioned the analogy of said Kelly's process in decarbonising iron to the process of decarbonising blood in the human lungs.

The Doctor does not say, specifically, if he or any of the "company" went to see the process in operation.

Kelly obtained affidavits from another seventeen witnesses. Ten of these recorded their recollections of experiments conducted in 1847. Five described the 1851 work. Two knew of or had seen both. One of the last group was John B. Evans who became forge manager of Kelly's Union Forge, a few miles from Suwanee. This evidence is of interest since a man in his position should have been in a position to tell something about the results of Kelly's operations in terms of usable metal. Unfortunately, he limits himself to a comment on the metal which had chilled around a tuyère which had been sent back to the Forge ("it was partly malleable and partly refined pig-iron") and to an account of a conversation with others who had worked some of Kelly's "good wrought iron" made by the new process.

Only one of the witnesses (William Soden) makes a reference to the phenomenon which is an accompaniment of the blowing of a converter: the prolonged and violent emission of sparks and flames which startled Bessemer in his first use of the process[99] and which still provides an exciting, if not awe-inspiring, interlude in a visit to a steel mill. Soden refers, without much excitement, to a boiling commotion, but the results of Kelly's "air-boiling" were, evidently, not such as to impress the rest of those who claimed to have seen his furnace in operation. Only five of the total of eighteen of the witnesses say that they witnessed the operations. Soden, incidentally, knew of seven different "air-boiling" furnaces, some with four and some with eight tuyères, but he also neglected to report on the use of the metal.

As is well known, Kelly satisfied the Acting Commissioner that he had "made this invention and showed it by drawings and experiment as early as 1847," and he was awarded priority by the Acting Commissioner's decision of April 13, 1857, and U.S. Patent 17628 was granted him as of June 23, 1857. The *Scientific American* sympathized with Bessemer's realization that his American patent was "of no more value to him than so much waste paper" but took the opportunity of chastising Kelly for his negligence in not securing a patent at a much earlier date and complained of a patent system which did not require an inventor to make known his discovery promptly. The journal advocated a "certain fixed time" after which such an inventor "should not be allowed to subvert a patent granted to another who has taken proper measures to put the public in possession of the invention."[100]

Little authentic is known about Kelly's activities following the grant of his patent. His biographer[101] does not document his statements, many of which appear to be based on the recollections of members of Kelly's family, and it is difficult to reconcile some of them with what few facts are available. Kelly's own account of his invention,[102] itself undated, asserts that he could "refine fifteen hundredweight of metal in from five to ten minutes," his furnace "supplying a cheap method of making run-out metal" so that "after trying it a few days we entirely dispensed with the old and troublesome run-out fires."[103] This statement suggests that Kelly's method was intended to do just this; and it is not without interest to note that several of his witnesses in the Interference proceedings, refer to bringing the metal "to nature," a term often used in connection with the finery furnace. If this is so, his assumption that he had anticipated Bessemer was based on a misapprehension of what the latter was intending to do, that is, to make steel.

This statement leaves the reader under the impression that the process was in successful use. It is to be contrasted with the statement quoted above (page 43), dated September 1856, when the process had, clearly, not been perfected. In this connection, it should be noted that in the report on the Suwanee Iron Works, included in *The iron manufacturer's guide*,[104] it is stated that "It is at this furnace that Mr. Kelly's process for refining iron in the hearth has been most fully experimented upon."

A major financial crisis affected United States business in the fall of 1857. It began in the first week of October and by October 31 the *Economist* (London) reported that the banks of the United States had "almost universally suspended specie payment."[105] Kelly was involved in this crisis and his plant was closed down. According to Swank,[106] some experiments were made to adapt Kelly's process to need of rolling mills at the Cambria Iron Works in 1857 and 1858, Kelly himself being at Johnstown, at least in June 1858. That the experiments were not particularly successful is suggested by the lack of any American contributions to the correspondence in the English technical journals. Kelly was not mentioned as having done more than interfere with Bessemer's first patent application. The success of the latter in obtaining patents[107] in the United States in November 1856, covering "the conversion of molten crude iron ... into steel or malleable iron, without the use of fuel ..." also escaped the attention of both English and American writers.

It was not until 1861 that the question arose as to what happened to Kelly's process. The occasion was the publication of an account of Bessemer's paper at the Sheffield meeting of the (British) Society of Mechanical Engineers on August 1, 1861. Accepting the evidence of "the complete industrial success" of Bessemer's process, the *Scientific*

American[108] asked: "Would not some of our enterprising manufacturers make a good operation by getting hold of the [Kelly] patent and starting the manufacture of steel in this country?"

There was no response to this rhetorical question, but a further inquiry as to whether the Kelly patent "could be bought"[109] elicited a response from Kelly. Writing from Hammondsville, Ohio, Kelly[110] said, in part:

I would say that the New England states and New York would be sold at a fair rate.... I removed from Kentucky about three years ago, and now reside at New Salisbury about three miles from Hammondsville and sixty miles from Pittsburg. Accept my thanks for your kind efforts in endeavoring to draw the attention of the community to the advantages of my process.

This letter suggests that the Kelly process had been dormant since 1858. Whether or not as a result of the publication of this letter, interest was resumed in Kelly's experiments. Captain Eber Brock Ward of Detroit and Z. S. Durfee of New Bedford, Massachusetts, obtained control of Kelly's patent. Durfee himself went to England in the fall of 1861 in an attempt to secure a license from Bessemer. He returned to the United States in the early fall of 1862, assuming that he was the only "citizen of the United States" who had even seen the Bessemer apparatus.[111]

In June, 1862, W. F. Durfee, a cousin of Z. S. Durfee, was asked by Ward to report on Kelly's process. The report[112] was unfavorable. "The description of [the apparatus] used by Mr. Kelly at his abandoned works in Kentucky satisfied me that it was not suited for an experiment on so large a scale as was contemplated at Wyandotte [Detroit]." Since it was "confidently expected that Z. S. Durfee would be successful in his efforts to purchase [Bessemer's patents], it was thought only to be anticipating the acquisition of property rights ... to use such of his inventions as best suited the purpose in view."

Thus the first "Bessemer" plant in the United States came into being without benefit of a license and supported only by a patent "not suited" for a large experiment. Kelly seems to have had no part in these developments. They took some time to come to formation. Although the converter was ready by September 1862, the blowing engine was not completed until the spring of 1864 and the first "blow" successfully made in 1864. It may be no more than a coincidence that the start of production seems to have been impossible before the arrival in this country of a young man, L. M. Hart, who had been trained in Bessemer operations at the plant of the Jackson Brothers at St. Seurin (near Bordeaux) France. The Jacksons had become Bessemer's partners in respect of the French rights; and the recruitment of Hart suggests the possibility that it was from this French source that Z. S. Durfee obtained his initial technical data on the operation of the Bessemer process.[113]

During the organization of the plant at Wyandotte, Kelly was called back to Cambria, probably by Daniel J. Morrell, who, later, became a partner with Ward and Z. S. Durfee in the formation of the Kelly Pneumatic Process Company.[114] We learn from John E. Fry,[115] the iron moulder who was assigned to help Kelly, that —

in 1862 Mr. Kelly returned to Johnstown for a crucial, and as it turned out, a final series of experiments by him with a rotative [Bessemer converter] *made abroad and imported for his purpose.* This converter embodied in its materials and construction several of Mr. Bessemer's patented factors, of which, up to the close of Mr. Kelly's experiments above noted, he seemed to have no knowledge or conception. And it was as late as on the occasion of his return in 1862, to operate the experimental Bessemer converter, that he first recognized, by its adoption, the necessity for or the importance of any after treatment of, or additions required by the blown metal to convert it into steel.

Fry later asserted[116] that Kelly's experiments in 1862 were simply attempts to copy Bessemer's methods. (The possibility is under investigation that the so-called "pioneer converter" now on loan to the U.S. National Museum from the Bethlehem Steel Company, is the converter referred to by Fry.)

William Kelly, in effect, disappeared from the record until 1871 when he applied for an extension of his patent of June 23, 1857. The application was opposed (by whom, the record does not state) on the grounds that the invention was not novel when it was originally issued, and that it would be against the public interest to extend its term. The Commissioner ruled that,[117] on the first question, it was settled practice of the Patent Office not to reconsider former decisions on questions of fact; the novelty of Kelly's invention had been re-examined when the patent was reissued in November 1857. Testimony showed that the patent was very valuable; and that Kelly "had been untiring in his efforts to introduce it into use but the opposition of iron manufacturers and the amount of capital required prevented him from receiving anything from his patent until within very few years past." Kelly's expenditures were shown to have amounted to $11,500, whereas he had received only $2,400. Since no evidence was filed in support of the public interest aspect of the case, the Commissioner found no substantial reason for denying the extension; indeed "very few patentees are able to present so strong grounds for extension as the applicant in the case."

In a similar application in the previous year, Bessemer had failed to win an extension of his U.S. patent 16082, of November 11, 1856, for the sole reason that his British patent with which it had been made co-terminal had duly expired at the end of its fourteen years of life, and it would have been inequitable to give Bessemer protection in the United States while British iron-masters were not under similar restraint. But if it had not been for this consideration, Bessemer "would be justly entitled to what he asks on this occasion." The Commissioner[118] observed: "It may be questioned whether [Bessemer] was first to discover the principle upon which his process was founded. But we owe its reduction to practice to his untiring industry and perseverance, his superior skill and science and his great outlay."

Conclusions

Martien was probably never a serious contender for the honor of discovering the atmospheric process of making steel. In the present state of the record, it is not an unreasonable assumption that his patent was never seriously exploited and that the Ebbw Vale Iron Works hoped to use it, in conjunction with the Mushet patents, to upset Bessemer's patents.

The position of Mushet is not so clear, and it is hoped that further research can eventually throw a clearer light on his relationship with the Ebbw Vale Iron Works. It may well be that the "opinion of metallurgists in later years"[119] is sound, and that both Mushet and Bessemer had successfully worked at the same problem. The study of Mushet's letters to the technical press and of the attitude of the editors of those papers to Mushet suggests the possibility that he, too, was used by Ebbw Vale for the purposes of their attacks on Bessemer. Mushet admits that he was not a free agent in respect of these patents, and the failure of Ebbw Vale to ensure their full life under English patent law indicates clearly enough that by 1859 the firm had realized that their position was not strong enough to warrant a legal suit for infringement against Bessemer. Their purchase of the Uchatius process and their final attempt to develop Martien's ideas through the Parry patents, which exposed them to a very real risk of a suit by Bessemer, are also indications of the politics in the case. Mushet seems to have been a willing enough victim of Ebbw Vale's scheming. His letters show an almost presumptuous assumption of the mantle of his father; while his sometimes absurd claims to priority of invention (and demonstration) of practically every new idea in the manufacturing of iron and steel progressively reduced the respect for his name. Bessemer claims an impressive array of precedents for the use of manganese in steel making and, given his attitude to patents and his reliance on professional advice in this respect, he should perhaps, be given the benefit of the doubt. A dispassionate judgment would be that Bessemer owed more to the development work of his Swedish licensees than to Mushet.

Kelly's right to be adjudged the joint inventor of what is now often called the Kelly-Bessemer process is questionable.[120] Admittedly, he experimented in the treatment of molten metal with air blasts, but it is by no means clear, on the evidence, that he got beyond the experimental stage. It is certain that he never had the objective of making steel, which was Bessemer's primary aim. Nor is there evidence that his process was taken beyond the experimental stage by the Cambria Works. The rejection of his "apparatus" by W. F. Durfee must have been based, to some extent at least, upon the Johnstown trials. There are strong grounds then, for agreeing with one historian[121] who concludes:

The fact that Kelly was an American is evidently the principal reason why certain popular writers have made much of an invention that, had not Bessemer developed his process, would never have attracted notice. Kelly's patent proved very useful to industrial interests in this country as a bargaining weapon in negotiations with the Bessemer group for the exchange of patent rights.

Kelly's suggestion[122] that some British puddlers may have communicated his secret to Bessemer can, probably, never be verified. All that can be said is that Bessemer was not an ironman; his contacts with the iron trade were, so far as can be ascertained, nonexistent until he himself invaded Sheffield. So it is unlikely that such a secret would have been taken to him, even if he were a well-known inventor.

Footnotes

[1] From 1870 through 1907, "Bessemer" production accounted for not less than 50 percent of United States steel production. From 1880 through 1895, 80 percent of all steel came from this source: Historical Statistics of the United States 1789-1945 (Washington, U.S. Department of Commerce, Bureau of the Census, 1949), Tables J. 165-170 at p. 187.

[2] See especially material distributed by the American Iron and Steel Institute in connection with its celebration of the centennial of Steel: "Steel centennial (1957), press information," prepared by Hill and Knowlton, Inc., and released by the Institute as of May 1, 1957

[3] Holley's work is outside the scope of this paper. Belatedly, his biography is now being written. It can hardly fail to substantiate the contention that during his short life (1832-1882) Holley, who negotiated the purchase of the American rights to Bessemer's process, also adapted his methods to the American scene and laid a substantial part of the foundation for the modern American steel industry

[4] Andrew Ure, *Dictionary of arts, manufactures and mines*, New York, 1856, p. 735

[5] See abridgement of British patent 8021 of 1839 quoted by James S. Jeans, *Steel*, London, 1880, p. 28 ff. It is not clear that Heath was aware of the precise chemical effect of the use of manganese in this way

[6] *Mining Journal*, 1857, vol. 27, p. 465

[7] *Sir Henry Bessemer, F.R.S., an autobiography*, London, 1905, p. 332

[8]*Ibid.*, p. 59 ff

[9]*Ibid.*, p. 82

[10]*Ibid.*, p. 83

[11]*Ibid.*, p. 108 ff

[12]*Ibid.*, p. 141. Bessemer's assertion that he had approached "within measurable distance" of anticipating the Siemens-Martin process, made in a paper presented at a meeting of the American Society of Mechanical Engineers (*Transactions of the American Society of Mechanical Engineers*, 1897, vol. 28, p. 459), evoked strong criticism of Bessemer's lack of generosity (*ibid.*, p. 482). One commentator, friendly to Bessemer, put it that "Bessemer's relation to the open-hearth process was very much like Kelly's to the Bessemer process.... Although he was measurably near to the open-hearth process, he did not follow it up and make it a commercial success...." (*ibid.*, p. 491)

[13]British patent 2489, November 24, 1854

[14]Bessemer, *op. cit.* (footnote 7), p. 137 He received British patent 66, dated January 10, 1855

[15]See James W. Dredge, "Henry Bessemer 1813-1898," *Transactions of the American Society of Mechanical Engineers*, 1898, vol. 19, p. 911

[16]See U.S. Patent Office, Decision of Commissioner of Patents, dated April 13, 1857, in Kelly vs. Bessemer Interference. This is further discussed below (p. 42)

[17] Dredge, *op. cit.* (footnote 15), p. 912

[18] Bessemer's paper was reported in *The Times*, London, August 14, 1856. By the time the Transactions of the British Association were prepared for publication, the controversy aroused by Bessemer's claim to manufacture "malleable iron and steel without fuel" had broken out and it was decided not to report the paper. Dredge (*op. cit.*, footnote 15, p. 915) describes this decision as "sagacious."

[19] Bessemer, *op. cit.* (footnote 7), p. 164

[20] *The Times*, London, August 14, 1856

[21] David Mushet recognized that Bessemer's great feature was this effort to "raise the after processes ... to a level commensurate with the preceding case" (*Mining Journal*, 1856, p. 599)

[22] See *Mining Journal*, 1857, vol. 27, pp. 839 and 855. David Mushet withdrew from the discussion after 1858 and his relapse into obscurity is only broken by an appeal for funds for the family of Henry Cort. A biographer of the Mushets is of the opinion that Robert Mushet wrote these letters and obtained David's signature to them (Fred M. Osborn, *The story of the Mushets*, London, 1952, p. 44, footnote). The similarity in the style of the two brothers is extraordinary enough to support this idea. If this is so, Robert Mushet who disagreed with himself as "Sideros" was also in controversy with himself writing as "David."

[23] *Mining Journal*, 1856, vol. 26, p. 567

[24] *Ibid.*, pp. 631 and 647. The case of Martien will be discussed below (p. 36). David Mushet had overlooked Bessemer's patent of January 10, 1855

[25] *Mining Journal*, 1857, vol. 27, p. 723. Robert Mushet was a constant correspondent of the *Mining Journal* from 1848. The adoption of a pseudonym, peculiar apparently to 1857-1858 (see *Dictionary of national biography*, vol. 39, p. 429), enabled him to carry on two debates at a time and also to sing his own praises

[26] *Ibid.*, p. 823. Mushet's distinction between an inventor and a patentee is indicative of the disdain of a son of David Mushet for an amateur (see also p. 886)

[27] One William Green had commented extensively on David Mushet's early praise of the Bessemer process and on his sudden reversal in favor of Martien soon after Bessemer's British Association address (*Mechanics' Magazine*, 1856, vol. 65, p. 373 ff.). Green wrote from Caledonian Road, and the proximity to Baxter House, Bessemer's London headquarters, suggests the possibility that Green was writing for Bessemer

[28] *Mining Journal*, 1857, vol. 27, p. 764

[29] *Ibid.*, p. 764

[30] *Ibid.*, p.791

[31] *Ibid.*, p. 770 (italics supplied)

[32] *Ibid.*, p. 770

[33] *Ibid.*, p. 823

³⁴Bessemer, *op. cit.* (footnote 7), p. 169

³⁵*Mining Journal*, 1856, vol. 26, p. 631

³⁶James Renton's process (U.S. patent 8613, December 23, 1851) had been developed at Newark, New Jersey, in 1854. It was a modification of the puddling furnace, in which the ore and carbon were heated in tubs, utilizing the waste heat of the reverberatory furnace (see the *Mechanics' Magazine*, vol. 62, p. 246, 1855). Renton died at Newark in September 1856 (*Mechanics' Magazine*, 1856, vol. 65, p. 422)

³⁷*Mining Journal*, 1857, vol. 27, p. 193

³⁸British patent 2219, September 22, 1856

³⁹Joseph P. Lesley, *The iron manufacturer's guide*, New York, 1859, p. 34. Martien's name is spelled Marteen. A description of the furnace is given in *Scientific American* of February 11, 1854, (vol. 9, p. 169). In the patent interference proceedings referred to below, it was stated that the furnace was in successful operation in 1854

⁴⁰U.S. patent 16690, February 22, 1857. A correspondent of the *Mining Journal* (1858, vol. 28, p. 713) states that Martien had not returned to England by October 1858

⁴¹U.S. Patent Office, Decision of Commissioner of Patents, dated May 26, 1859 in the matter of interference between the application of James M. Quimby and others ... and of Joseph Martien

[42] J. S. Jeans, *op. cit.* (footnote 5), p. 108. The process is not mentioned by James M. Swank, *History of the manufacture of iron in all ages*, Philadelphia, American Iron and Steel Association, 1892

[43] *Mining Journal*, 1856, vol. 26, p. 707

[44] Bessemer, *op. cit.* (footnote 7), p. 290

[45] The American Iron and Steel Institute's "Steel centennial (1957) press information" (see footnote 2), includes a pamphlet, "Kelly lighted the fireworks ..." by Vaughn Shelton (New York, 1956), which asserts (p. 12) that Bessemer paid the renewal fee and became the owner of Mushet's "vital" patent

[46] Robert Mushet, *The Bessemer-Mushet process*, Cheltenham, 1883, p. 24; *The Engineer*, 1861, vol. 12, pp. 177 and 189

[47] *The Engineer*, 1862, vol. 14, p. 3. Bessemer, *op. cit.* (footnote 7), p. 296

[48] *Mining Journal*, 1864, vol. 34, p. 478

[49] *The Engineer*, 1861, vol. 12, p. 189

[50] *Ibid.*, p. 78

[51] Mushet, *op. cit.* (footnote 46), p. 9

[52] *Ibid.*, p. 25

[53] *Mining Journal*, 1857, vol. 27, p. 755

[54] Mushet, *op. cit.* (footnote 46), p. 28. The Uchatius process became the "You-cheat-us" process to Mushet (*Mining Journal*, 1858, vol. 28, p. 34)

[55] *Mining Journal*, 1857, vol. 27, p. 755 (italics supplied)

[56] See footnote 22

[57] *Mining Journal*, 1856, vol. 26, pp. 583, 631

[58] October 17, 1857, writing as "Sideros" (*Mining Journal*, 1857, vol. 27, p. 723)

[59] *Mining Journal*, 1857, vol. 27, p. 871, and 1858, vol. 28, p. 12

[60] *Ibid.* (1858), p. 34

[61] Mushet, *op. cit.* (footnote 46), p. 12. The phrase quoted is typical of Mushet's style

[62] Bessemer, *op. cit.* (footnote 7), pp. 161 ff. and 256 ff

[63] *Ibid.*, p. 171

[64] This enterprise, started in conjunction with Galloway's of Manchester, one of the firms licensed by Bessemer to make his equipment, was under way by April 1858 (see *Mining Journal*, 1858, vol. 28, p. 259)

[65]*Mining Journal*, 1858, vol. 28, p. 696. Mushet commented (p. 713) that he had done the same thing "eighteen months ago."

[66]Swank, *op. cit.* (footnote 42), p. 405

[67]*The Engineer*, 1859, vol. 7, p. 350

[68]*Mining Journal*, 1859, vol. 29, pp. 396 and 401. The price quotation was continued until April 1865

[69]*The Engineer*, 1859, vol. 7, p. 437

[70]Jeans, *op. cit.* (footnote 5), p. 349 refers to the hematite ores of Lancashire and Cumberland as "the ores hitherto almost exclusively used in the Bessemer process."

A definitive account of the Swedish development of the Bessemer process, leading to a well-documented claim that the first practical realization of the process was achieved in Sweden in July 1858, was recently published (Per Carlberg, "Early Production of Bessemer Steel at Edsken," *Journal of the Iron and Steel Institute, Great Britain*, July 1958, vol. 189, p. 201)

[71]*The Engineer*, 1859, vol. 7, p. 314. Bessemer's intention to present his paper had been announced in April

[72]*Mining Journal*, 1859, vol. 29, p. 539 and 640. Another Mushet patent is described as so much like Uchatius' process that it would seem to be almost unpatentable

⁷³See Jeans, *op. cit.* (footnote 5), p. 532

⁷⁴*The Engineer*, 1859, vol. 8, p. 13 (italics supplied). It is noted that Mushet's American patent (17389, of May 26, 1857) prefers the use of iron "as free as possible from Sulphur and Phosphorous."

⁷⁵*The Engineer*, 1860, vol. 9, pp. 366, 416, and *passim*

⁷⁶*The Engineer*, 1861, vol. 11, pp. 189, 202, 290, 304

⁷⁷*The Engineer*, 1861, vol. 12, p. 10

⁷⁸*Ibid.*, p. 63

⁷⁹*Ibid.*, pp. 78 and 177

⁸⁰*Ibid.*, p. 208. There is an intriguing reference in this editorial to an interference on behalf of Martien against a Bessemer application for a U.S. patent. No dates are given and the case has not been located in the record of U.S. Patent Commissioner's decision

⁸¹*Ibid.*, p. 254

⁸²U.S. patent 17389, dated May 26, 1857. The patent was not renewed when application was made in 1870, on the grounds that the original patent had been made co-terminal with the British patent. The latter had been abandoned "by Mushet's own fault" so that no right existed to an American renewal (U.S. Patent Office, Decision of Commissioner of Patents, dated September 19, 1870)

[83]See below, p. 45. The exact date of the purchase of Mushet's patent is not known

[84]*Engineering*, 1882, vol. 33, p. 114. The deal was completed in 1863

[85]*The Engineer*, 1864, vol. 18, pp. 405, 406

[86]*Mining Journal*, 1864, vol. 34, pp. 77 and 94 (italics supplied). It has not yet been possible to ascertain if this company was successful. Mushet writes from this time on from Cheltenham, where the company had its offices. Research continues in this interesting aspect of his career

[87]*Mining Engineer*, 1865, vol. 35, p. 86

[88]*The Engineer*, 1865, vol. 20, p. 174

[89]*Mechanics' Magazine*, 1866, vol. 16, p. 147

[90]Bessemer, *op. cit.* (footnote 7), p. 294

[91]*Ibid.*

[92]See Fred M. Osborn, *The story of the Mushets*, London, 1852

[93]*Journal of the Iron and Steel Institute*, 1876, p. 3

[94]Robert Mushet, *The Bessemer-Mushet process*, Cheltenham, 1883

[95] *Scientific American*, 1856, vol. 12, p. 6

[96] U.S. patent 17389, dated May 26, 1857. Martien's U.S. patent was granted as 16690, dated February 24, 1857

[97] *Scientific American*, 1856, vol. 12, p. 43, Kelly's suggestion of piracy of his ideas was later enlarged upon by his biographer John Newton Boucher, *William Kelly: A true history of the so-called Bessemer process*, Greensburg, Pennsylvania, 1924

[98] *Ibid.*, p. 82. Kelly's notice of his intention to take testimony was addressed to Bessemer on January 12, 1857. See papers on "Interference, William Kelly vs. Henry Bessemer Decision April 13, 1857." U.S. Patent Office Records. Quotations below are from this file, which is now permanently preserved in the library of the U.S. Patent Office

[99] Bessemer, *op. cit.* (footnote 7), p. 144

[100] *Scientific American*, 1857, vol. 12, p. 341

[101] Boucher, *op. cit.* (footnote 97)

[102] U.S. Bureau of the Census, *Report on the manufacturers of the United States at the tenth census (June 1, 1880) ..., Manufacture of iron and steel*, report prepared by James M. Swank, special agent, Washington, 1883, p. 124. Mr. Swank was secretary of the American Iron and Steel Association. This material was included in his *History of the manufacture of iron in all ages*, Philadelphia, 1892, p. 397

[103] *Ibid.*, p. 125. The run-out fire (or "finery" fire) was a charcoal fire "into which pig-iron, having been melted and partially refined in

one fire, was run and further refined to convert it to wrought iron by the Lancashire hearth process," according to A. K. Osborn, *An encyclopaedia of the iron and steel industry*, New York, 1956

[104]J. P. Lesley, *op. cit.* (footnote 39), p. 129. The preface is dated April 6, 1859. The data was largely collected by Joseph Lesley of Philadelphia, brother of the author, during a tour of several months. Since Suwanee production is given for 44 weeks only of 1857 (*i.e.*, through November 4 or 5, 1857) it is concluded that Lesley's visit was in the last few weeks of 1857

[105]*Economist* (London), 1857, vol. 15, pp. 1129, 1209

[106]Swank, *op. cit.* (footnote 42), p. 125. John Fritz, in his *Autobiography* (New York, 1912, p. 162), refers to experiments during his time at Johnstown, *i.e.*, between June 1854 and July 1860. *The iron manufacturer's guide* (see footnote 104) also refers to Kelly's process as having "just been tried with great success" at Cambria

[107]U.S. patents 16082, dated November 11, 1856, and 16083, dated November 18, 1856. Bessemer's unsuccessful application corresponded with his British patent 2321, of 1855 (see footnote 98)

[108]*Scientific American*, 1861, new ser., vol. 5, pp. 148-153

[109]*Ibid.*, p. 310

[110]*Ibid.*, p. 343

[111]His claim is somewhat doubtful. Alexander Lyman Holley, who was later to be responsible for the design of most of the first Bessemer plants in the United States had been in England in 1859, 1860,

and 1862. In view of his interest in ordnance and armor, it is unlikely that Bessemer could have escaped his alert observation. His first visit specifically in connection with the Bessemer process appears to have been in 1863, but he is said to have begun to interest financiers and ironmasters in the Bessemer process after his visit in 1862 (*Engineering*, 1882, vol. 33, p. 115)

[112]W. F. Durfee: "An account of the experimental steel works at Wyandotte, Michigan," *Transactions of the American Society of Mechanical Engineers*, 1884, vol. 6, p. 40 ff

[113]Research in the French sources continues. The arrival of L. M. Hart at Boston is recorded as of April 1, 1864, his ship being the SS *Africa* out of Liverpool, England (Archives of the United States, card index of passenger arrivals 1849-1891 list No. 39)

[114]Swank, *op. cit.* (footnote 42), p. 409

[115]*Johnstown Daily Democrat*, souvenir edition, autumn 1894 (italics supplied). Mr. Fry was at the Cambria Iron Works from 1858 until after 1882

[116]*Engineering*, 1896, vol. 61, p. 615

[117]See U.S. Patent Office, Decision of Commissioner of Patents, dated June 15, 1871

[118]U.S. Patent Office, Decision of Commissioner of Patents dated February 12, 1870

[119]William T. Jeans, *The creators of the age of steel*, London, 1884

[120]Bessemer dealt with Kelly's claim to priority in a letter to *Engineering*, 1896, vol. 61, p. 367

[121]Louis C. Hunter, "The heavy industries since 1860," in H. F. Williamson (editor), *The growth of the American economy*, New York, 1944, p. 469

[122]Later developed into a dramatic story by Boucher, *op. cit.* (footnote 97)

www.ingramcontent.com/pod-product-compliance
Lightning Source LLC
Chambersburg PA
CBHW030511220526
45464CB00006B/2748